Be a Food SCIENTIST

• Question • Experiment • Discover

By Ruth Owen

Ruby Tuesday Books

Published in 2024 by Ruby Tuesday Books Ltd.

Copyright © 2024 Ruby Tuesday Books Ltd.

All rights reserved. No part of this publication may be reproduced in whole or in part, stored in any retrieval system, or transmitted in any form or by any means, electronic, mechanical, photocopying, recording, or otherwise, without written permission from the publisher.

Editor: Mark J. Sachner
Design: Tammy West
Production: John Lingham

Photo credits:
Alamy: 12BL (Simon Curtis), 25T (zzzdim); Ruby Tuesday Books: 3, 13, 14–15, 26–27; Shutterstock: Cover TL (EvgeniiAnd), Cover TC (Nedrofly), Cover TR (An Nguyen), Cover BL (Shulevskyy Volodymyr), Cover BC (New Africa), Cover BR (mikeledray), 1 (Pixel-Shot), 2, 4, 4TR (Kung37), 5TL (Nedrofly), 5C (Pixel-Shot), 6–7, 8–9, 10T (AYO Production), 10C, 10B (Clara Bastian), 11, 11C (Queenmoonlite Studio), 12TR (Prostock-studio), 13, 14TL (Gus Andrade), 16–17, 18–19, 18C (PeopleImages.com–Yuri A), 19BL (MIA Studio), 20–21, 22–23, 24T (wavebreakmedia), 24B (Marcos Castillo), 25B (OlgaBombologna), 26–27, 28–29, 30–31, 30CR (voronaman), 30BR (Pixel-Shot), 31CL (lightwavemedia).

ISBN 978-1-78856-432-8

Printed in Poland by L&C Printing Group

www.rubytuesdaybooks.com

Note From the Publisher
Neither the publisher nor the author can accept legal responsibility or liability for any loss, harm or injury that may come about from following the instructions in this book. All activities should be carried out with adult guidance and supervision. Some activities involve using kitchen equipment and touching food stuffs that could contain allergens. Children should be accompanied at all times. It is the parent's or carer's responsibility to ensure their child is safe.

Contents

There's Science on Your Plate!..................4

Solid, Liquid, Solid? 6

Make Rainbow Milk with Science............ 8

Let's Make Butter! 10

WOW! I Can Make Plastic.................... 12

Colour-Changing Rainbow Noodles..... 14

The Science of a Cake 16

More Cake Science 18

Veggie Science 20

Let's Investigate Seeds....................22

Meet the Food Inventors......................24

Be an Ice Cream Inventor 26

Let's Talk Food 28
Glossary....................................30
Index 32

There's Science on Your Plate!

You eat food every day. But did you know there is science in the foods you eat?

It's science that turns these **ingredients** into a fluffy sponge cake.

When lemon juice turns noodles pink, that's because of science!

And did you know that your favourite ice cream flavour or chocolate bar was invented by a scientist?

Some of the experiments in this book explain how foods are made.

Others show us science in action by using foods as our materials.

Always make sure an adult is nearby when you are working with food, knives, hot saucepans and other kitchen equipment.

Ready to have some fun with food? It's time to . . .

. . . be a **Food Scientist**.

Solid, Liquid, Solid?

Everything around us is made of **matter**. There are three states, or types, of matter — **solids**, **liquids** and gases.

Solids

- A solid has a shape.
- You can hold a solid.
- A solid can be cut into smaller pieces.

Liquids

- A liquid has no shape.
- A liquid is difficult to hold.
- A liquid flows and pours.

Gases

- A gas has no shape.
- Gases float in the air all around us.
- Most gases are invisible.

An eggshell is solid. But the egg inside is a thick liquid. Once the egg is cooked, it turns solid.

Solid eggshell

Liquid egg

Solid cooked egg

Can solids change to liquids?

Let's investigate some food items. Will heat **melt** them and make them liquid? And then will they turn solid again once they are cooled?

You will need:
- A notebook and pencil
- 6 items for testing (see right)
- 6 small foil dishes
- A baking tray
- An adult helper and hot water

Ice cubes   Butter   Grated cheese
Chocolate   Jelly cubes   Scrambled egg

1) Draw this chart in your notebook.

	Ice	Butter	Grated cheese	Chocolate	Jelly	Scrambled egg
PREDICTION						
Melts when heated						
Becomes solid again when cooled						
RESULTS						
Melts when heated						
Becomes solid again when cooled						

2) Look at each food. Predict if it will melt by writing YES or NO on your chart. Then predict if it will become solid again once cooled.

3) Put a small amount of each food into a foil dish.

4) Ask your adult helper to carefully pour hot water from a kettle into the baking tray. Then ask them to float your foil dishes on the hot water.

5) As the foods heat up, observe what happens and record your results.

6) Allow the hot water to cool. Observe any changes as the food items cool down, and record your results.

Did your predictions match what happened?

To find answers and more information, turn to page 28.

Make Rainbow Milk with Science

When we spill liquids such as water, milk and juice, they form drops and puddles. Why?

The liquids are made of tiny parts called **molecules**.

The molecules stick together tightly.

Drop

On the liquid's surface, the molecules hold onto each other extra tight.

This is called surface tension.

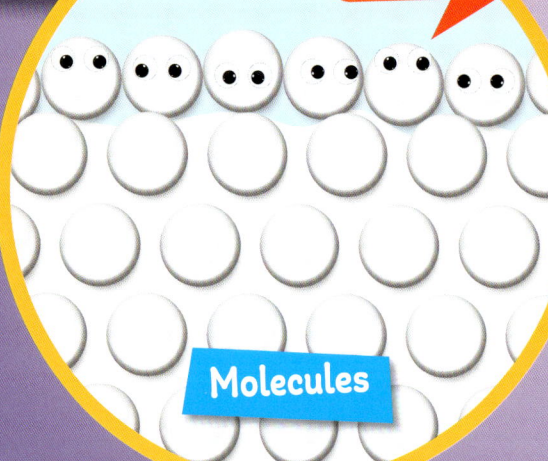

Hold on tight, everyone!

Molecules

Molecules on a liquid's surface stick to the ones below them and on either side.

Let's watch liquid molecules make a rainbow!

This activity is a fun and colourful way to see molecules and surface tension in action.

You will need:
- A shallow dish
- 1 pint (568 ml) of full fat milk
- Food colouring
- Cotton buds
- Washing-up liquid

1) Pour milk into the dish so the bottom is covered.

2) Add one drop of each food colouring to the milk's surface in a line.

3) Gently touch the milk with a cotton bud close to the colours.

What do you observe is happening?

4) Take a new cotton bud and dip it in washing-up liquid. Now touch the milk in the same spot as before.

What happens now? What do you think the milk molecules are doing?
(Don't drink the milk as it contains washing-up liquid.)

Try the activity again with the colours in different places, or try touching the milk in different places. Have fun experimenting!

To find answers and more information, turn to page 28.

Let's Make Butter!

We drink milk and pour it onto our cereal.

This super food is filled with **nutrients** that keep our bones, muscles and teeth healthy.

But did you know that **cream**, butter, cheese and yoghurt are all made from milk?

Cream

Butter

Cheese

Yoghurt

Milk comes from cows and goats. It can also be made from plants.

Oat milk and soya milk are types of plant milks. Oats or soya beans are crushed and mixed with water to make a milky drink.

Cream is a thick liquid that contains lots of **fat**. It is used to make butter.

So how does liquid cream become solid butter? Let's investigate!

You will need:
- A jar with a tightly sealed lid
- A cup of double cream
- A pinch of salt
- Some helpers, including an adult for extra power!
- A phone or watch for timing

1) Pour the cream into the container. Add the salt.

2) Put on the lid and check that it is tightly sealed.

3) Now start shaking the container — hard! Take turns shaking with your helpers.

Team shake!

After 5 minutes, check the container. What do you observe is happening?

4) You will need to shake the cream for up to 10 minutes. A clump of solid butter will form.

5) Pour away the liquid in the container. Squeeze your lump of butter to remove any liquid. Your butter is now ready to eat.

What do you think has happened to change the liquid to a solid?

To find answers and more information, turn to page 29.

WOW! I Can Make Plastic

When milk and vinegar are mixed, a **chemical reaction** happens. The liquid mixture becomes a type of **plastic** called casein.

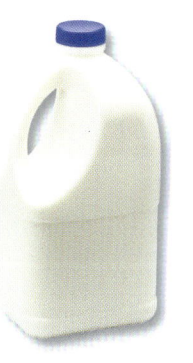    

Milk + **Vinegar** = **Plastic**

The chemical reaction makes some of the molecules in the milk change.

Scientists first developed and made casein in the early 1900s.

Casein necklace

Casein hair comb

The plastic is also known as Galalith.

Casein was used to make small items such as buttons, jewellery, pens and hairbrushes.

It's time to be a food scientist and turn liquid milk into solid plastic!

You will need:
- Measuring cups and spoons
- Cow or goat's milk
- A small saucepan
- An adult helper
- A small bowl
- Food colouring
- A spoon
- Distilled white vinegar
- A sieve
- Kitchen towels
- Mini cookie cutters

1) Pour 1 cup of milk into a saucepan. Ask an adult to help you heat the milk until it is warm and starting to steam, but not hot and bubbling.

2) Pour the warm milk into the bowl. If you wish to colour your plastic, stir in food colouring now.

3) Add 4 teaspoons of vinegar to the bowl. Very gently stir the mixture for 15 seconds. The mixture will form lumps called curds.

Curds

4) Over a sink, pour the mixture into a sieve. Press the curds with the spoon to help remove the liquid.

5) Spoon the curds onto kitchen towels. Fold the kitchen towels around the curds. Gently press to flatten the curds and remove any remaining liquid.

6) Now you can cut shapes from your soft plastic with cookie cutters.

7) Place your shapes on kitchen towels to dry. After 48 hours they will turn hard.

To find answers and more information, turn to page 29.

Colour-Changing Rainbow Noodles

Did you know you can make a salad with noodles that change colour?

It sounds like magic — but it's actually science. Let's investigate!

You will need:
- Water
- A saucepan
- ¼ red cabbage cut into chunks
- An adult helper
- A phone or watch for timing
- A trivet
- Tongs
- 1 or 2 nests thin white noodles
- A colander
- A bowl
- Your choice of salad leaves and vegetables
- A lemon
- A knife and cutting board
- Your choice of dressing

Cabbage

1) Put 2 cups of water into a saucepan. Add the pieces of red cabbage.

2) Ask your adult helper to boil the water and cabbage for 5 minutes. Remove the pan from the cooker and place on a heat-proof trivet.

3) Carefully remove the red cabbage pieces from the water with the tongs.

4) Now place the noodles into the hot water. Leave them to soak for 10 minutes.

What do you observe is happening to the noodles?

BIG SCIENCE ALERT!

A red cabbage contains chemicals called anthocyanins.

The chemicals will dye, or colour, your noodles purple.

Red cabbages, red and black grapes, berries, cherries and plums get their colour from anthocyanins. These chemicals are not just colourful, they are good for us too.

5) Place a colander in the sink. Carefully empty the water and noodles into the colander, and shake to remove the water.

6) Place salad leaves in a bowl. Then using the tongs, put the noodles on top.

7) Now carefully cut a lemon in half. Squeeze juice over the purple noodles.

What do you observe is happening?

BIG SCIENCE ALERT!

Lemon juice contains a chemical called acid.

It creates a chemical reaction that makes the purple anthocyanin change to pink.

8) Finish your noodle salad by adding crunchy raw vegetables and your favourite dressing.

Enjoy your healthy, science-packed salad!

The Science of a Cake

Did you know that baking a cake is like a science experiment? Let's investigate the changes and chemical reactions that happen when you combine baking ingredients and heat!

By following this cake recipe, you will turn butter, sugar, eggs and flour into something new.

You will need:
- Kitchen scales
- 4 eggs
- Soft butter
- Caster sugar
- Self-raising flour
- 2 17-cm cake tins
- A mixing bowl
- A whisk, wooden spoon or electric mixer
- Oven gloves
- An adult helper
- A phone, watch or kitchen timer
- A cooling rack
- Jam, cream and icing sugar

1) Begin by weighing the eggs. These eggs weigh 290g.

2) Now measure out the same weight of butter, sugar and flour.

3) Ask an adult to help you heat the oven to 180°C.

4) Grease the cake tins with a little butter to keep the sponges from sticking to the tins.

5) Put the butter and sugar into the mixing bowl. Beat with a whisk, wooden spoon or electric mixer until the mixture is light and fluffy.

The Science Stuff
As you beat the butter and sugar, tiny air bubbles get trapped in the mixture.

6) Add the eggs one at a time, beating them into the mixture. Add a teaspoon of vanilla extract for extra flavour.

The Science Stuff
One of the jobs of the liquid eggs is to mix all the ingredients together.

7) Now, add the flour and beat until all the ingredients form a smooth, thick mixture. (Turn to page 18.)

The Science Stuff
When flour is added to the wet mixture, substances in the flour join together. They form stretchy stuff called gluten.

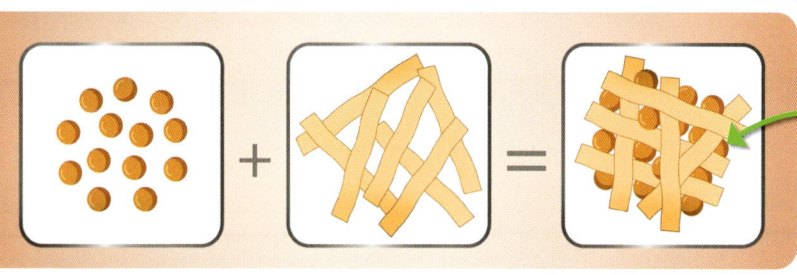

Gluten is like a web that holds a cake together.

More Cake Science

8) Divide the mixture equally between the two cake tins.

9) Ask an adult to help you put the cakes into the oven. Bake them for 25 minutes.

Now even more changes will happen!

The Science Stuff
The heat in the oven makes the trapped air bubbles in the mixture expand, or get bigger. This makes the cakes rise — just like blowing up a balloon.

Solid sponge

The Science Stuff
An ingredient called baking powder in self-raising flour releases carbon dioxide gas. This is the gas that makes fizzy drinks bubbly. The gas bubbles help the cake rise.

A cake rising

As the mixture gets hot, the eggs turn solid — just as they do when we boil or fry them.

Together, the eggs and gluten from the flour make the liquid mixture change to solid sponge.

Close-up picture of sugar grains

The Science Stuff
The tiny grains of solid sugar melt. They spread sweetness through the cake and give it a golden brown colour.

10) Ask your adult helper to remove the cakes from the oven. Place them on a cooling rack, and allow them to cool.

11) Spread a thick layer of jam and cream on top of one cake. Place the other cake on top. Dust with some icing sugar. This type of cake is called a Victoria sponge.

Chemistry is a type of science that studies what things are made of and how they can change. When we bake a cake, we are doing chemistry.

The Science Stuff
Baking a cake is a forever change. The ingredients can never go back to the materials they were before!

How do we know our cake is a solid? Because it has a shape.

And because we can cut it into smaller pieces — YUMMY!

19

Veggie Science

Many of the foods we eat come from plants. We eat roots, stems and leaves.

Carrot plant • Beetroot plant • Radish plant • Onion plant
leaf — Root — leaf — Bulb

When we eat a carrot, beetroot or a radish, we are eating the plant's root.

The part of an onion that we eat is called a bulb. It grows under the ground.

Lettuce plant

Cabbage plant

A lettuce is a plant, and so is a cabbage.

We eat the leaves of lettuce and cabbage plants.

Look closely at some broccoli. How would you describe the parts we eat?

What plant parts do you think they are? Let's investigate!

Broccoli is a vegetable that's packed with nutrients.

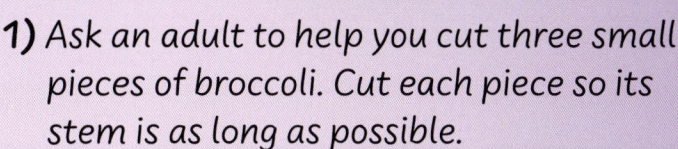

You will need:
- Some fresh broccoli
- An adult helper
- A knife and cutting board
- 3 glasses
- Water
- A notebook and pencil
- A magnifying glass or hand lens

1) Ask an adult to help you cut three small pieces of broccoli. Cut each piece so its stem is as long as possible.

2) Fill the glasses with water almost to the top.

3) Place a piece of broccoli into each glass. Stand the glasses on a warm, sunny windowsill.

4) What do you think will happen to the broccoli? Write your predictions in a notebook.

5) Check your broccoli every other day. Freshen the water if needed. Look for changes and record them in your notebook.

What do you observe is happening to the broccoli? Does it match your predictions?

Use a magnifying glass or hand lens to take a closer look!

To find answers and more information, turn to page 29.

Let's Investigate Seeds

Many of the plant foods we eat are **seeds**.

A seed is a tiny part of a plant that can grow into a new plant.

Bagel

Peanut butter

Carefully examine all the pictures.

Porridge   Baked beans   Baked potato

Pasta

Corn on the cob

Peas

Walnuts

Which of the pictures do you think show seeds we eat, or foods that are made from seeds?

Rice

Dead sunflower

To find answers and more information, turn to page 29.

Meet the Food Inventors

When you try a new breakfast cereal, or eat some delicious chocolate, do you ever wonder — who invented this food?

The answer is scientists called food technologists.

Their job is to invent new flavours and types of food for stores to sell.

Testing a new soup recipe

Have you ever tried fried egg, macaroni cheese, squid or mint-flavoured crisps? They've all been invented by food technologists!

Food technologists mix together ingredients and think up recipes.

They get to taste their inventions, too!

Food technologists work in a factory or a **laboratory**.

Sometimes food technologists invent new types of packaging to keep food fresh.

They may try to use materials that are plastic-free and good for the planet.

This ice cream container is made of a plant called bamboo. It can be put on a compost heap, and it will rot away.

Now it's your turn to think and work like a scientist and be an ice cream inventor!

Be an Ice Cream Inventor

When a food technologist invents something new, they carefully write a recipe that another person can follow.

Let's do some icy science.

You will need:
- A notebook and pencil
- A mixing bowl
- 2 cups double cream
- An electric hand mixer
- 1 tin condensed milk
- 1 teaspoon vanilla extract
- A wooden spoon
- Small bowls and spoons
- Food colouring
- A freezer-safe container with a lid
- Your choice of treats and decorations

1) Begin by thinking up an ice cream idea. Think about colours, ingredients you can add and decorations. Write your ideas in your notebook.

 Chunks of chocolate

Cookies

 Popcorn

 Nuts

 Dried fruit

 Cereal

 Sprinkles

 Strawberries

 Marshmallows

 Crackers or pretzels

2) To make some ice cream, pour the double cream into a mixing bowl. Beat the cream until it's thick and forms soft peaks.

3) Add the condensed milk and vanilla extract to the bowl. Stir with a wooden spoon until the mixture is thick and smooth.

You are now ready to get creative!

You can divide your ice cream into separate bowls to add different colours.

4) Add food colouring to your ice cream. Carefully count how many drops you add. Mix in your choice of treats (from step 1), recording everything you do in your notebook.

5) Spoon your ice cream into the freezer-safe container.

We added our ice cream colour by colour.

Finally, add decorations.

Remember! You can add whatever treats and decorations you like to your ice cream. But record each ingredient and measure each spoonful.

You can swirl the ice cream, too!

6) Place the container in the freezer, and leave to freeze overnight.

7) Now carefully write out the recipe for making your ice cream. Then give your ice cream a name!

Great work! You've become a food technologist.

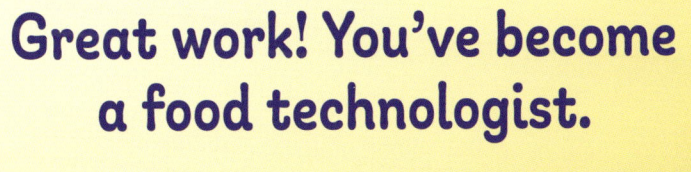

27

Let's Talk Food

Did you enjoy being a food scientist? Let's check out some answers and discover some more cool things about food science.

Page 7:
Some solids can become liquids and then change back again. But others can't.
When you heated the food items, the ice, butter and chocolate melted and became liquid. The cheese and jelly may have become softer and stickier, but they did not melt. The scrambled egg remained solid.

Melted chocolate

Melted butter

Once the foods cooled, the chocolate and butter became solid again. The water, from the ice, stayed liquid. But if you put that water in a freezer, it will freeze and become solid ice. Cheese needs a high heat to melt, such as under a hot grill. Once melted cheese cools, it's a rubbery solid, but it doesn't look the same as it did originally. Jelly melts or dissolves if it's put directly in hot water. Then, when the water cools down, the jelly becomes solid again. Once an egg is cooked, it can never go back to being a liquid raw egg.

Page 9:
When you touch the milk with the washing-up liquid, the colours instantly swirl, mix and make a rainbow. Why does this happen? That's some very BIG science!

At first, the surface tension of the milk keeps the drops of food colour floating on the milk's surface. However, the washing-up liquid changes the surface tension and makes the milk molecules move away from each other. This lets the food colouring spread out more easily through the milk.

There's also a second reason. Milk contains water and fat. The washing-up liquid molecules move around trying to find fat molecules to stick to. As they do this, they bump into the food colouring molecules, making them move and spread around, too.

Page 11:
As you shook the cream, a clump of butter gradually formed. Why?

Cream contains lots of molecules of fat. Shaking the cream made the fat molecules move around and bump into each other. The molecules started to stick together, making bigger and bigger fat molecules. Eventually, all the fat molecules clumped together and made a lump — butter!

The watery liquid that's left behind in the container is called buttermilk.

Page 13:
Why did the liquid milk become solid milk plastic?

Milk from animals contains molecules of a substance called casein. The vinegar makes the casein molecules change how they act. The molecules stick together and make curds that can be moulded into a plastic-like substance that becomes hard.

Broccoli flower

Page 21:
We eat a broccoli plant's stems. We also eat the frothy part that is lots of little flower buds. The buds open into yellow flowers that produce broccoli seeds.

Did your pieces of broccoli in water start to grow flowers?

Wheat

Pages 22-23:
All of the pictures (except for the potato) show seeds we eat!

The bagel and pasta are made from flour, which is made from the seeds of wheat plants. The bagel is topped with tiny poppy plant seeds. Porridge is made from oat plant seeds. And the rice we eat is rice plant seeds.

Beans, peas and sweetcorn are all seeds. A potato isn't a seed. It's a plant part called a tuber. A new potato plant can grow from a tuber. A walnut is a seed that can grow into a walnut tree. Sunflower seeds are used to make oil for cooking. They are also a crunchy, healthy snack.

Peanut butter is made from the seeds of peanut plants. A new plant can grow from the tiny, squiggly part inside a peanut.

Peanut shell

Peanut

A peanut plant grows from here.

Glossary

chemical reaction
A process in which substances mix together and a change happens. When baking soda and vinegar mix, it makes gas bubbles.

cream
A thick, fatty liquid. Milk from a cow is a mixture of water and fatty cream. If fresh milk is left to stand, the cream rises to the top.

fat
A nutrient that is found in foods such as butter, cheese, meat and nuts. We need to eat some fat to give us energy and keep us healthy.

ingredients
The different things that are used to make foods — for example, flour, butter, eggs and sugar are the ingredients for making a cake.

laboratory
A room or building where scientists use scientific equipment to do experiments.

liquid
A state of matter. Liquids flow and pour and do not have a shape.

matter
All the real stuff around us such as water, plants, books, food and our bodies.

melt
To turn from a solid into a liquid when heated.

molecules
Tiny building blocks that make up the things around us. Milk contains water molecules and fat molecules.

nutrient
A substance that a living thing needs to help it live and grow. Fruits and vegetables contain nutrients such as Vitamin D and Vitamin C.

plastic
A material that is usually made from oil, coal, chemicals and other materials. Plastic can be made into any shape or colour. It can be hard, soft and even see-through.

seed
A tiny part of a plant that contains all the material needed to grow a new plant.

solid
An object with a definite size and shape. Toys, this book and ice cubes are all solids.

Index

B
broccoli 21, 29
butter 7, 10–11, 12, 16–17, 28–29

C
cake 4, 16–17, 18–19
casein plastic 12, 29
chemical reactions 12, 15, 16
cream 10–11, 26, 28

E
eggs 6–7, 16–17, 18, 28

F
flour 16–17, 29
food technologists 4, 24–25, 26–27

G
gases 6, 18

I
ice cream 4, 25, 26–27

L
liquids 6–7, 8–9, 10–11, 12–13, 17, 18, 28–29

M
milk 8–9, 10, 12–13, 26, 28–29
molecules 8–9, 12, 28–29

N
noodles 4, 14–15
nutrients 10, 21

P
plastic 12–13, 25, 29

S
safety 5
seeds 22–23, 29
solids 6–7, 11, 12–13, 18–19, 28–29
sugar 16–17, 19

V
vegetables 14–15, 20–21, 29